USING ENGINEERING
TO FIGHT CLIMATE CHANGE

by Susan Wroble

WWW.FOCUSREADERS.COM

Copyright © 2023 by Focus Readers®, Lake Elmo, MN 55042. All rights reserved. No part of this book may be reproduced or utilized in any form or by any means without written permission from the publisher.

Focus Readers is distributed by North Star Editions:
sales@northstareditions.com | 888-417-0195

Produced for Focus Readers by Red Line Editorial.

Content Consultant: Stephen E. Still, PhD, Professor of Practice, Institute for Sustainable Transportation and Logistics, University at Buffalo

Photographs ©: Climeworks/Cover Images/AP Images, cover, 1, 21; David Burdick/NOAA, 4–5; Jonas Gratzer/Light Rocket/Getty Images, 6; Shutterstock Images, 8–9, 10, 14–15, 17, 22–23, 26 (background), 26 (aerosols), 26 (plane), 29; U.S. Department of Energy/Flickr, 12; Goddard/GSFC/NASA, 19; Michael Bell/The Canadian Press/AP Images, 25

Library of Congress Cataloging-in-Publication Data
Library of Congress Cataloging-in-Publication Data is available on the Library of Congress website.

ISBN
978-1-63739-276-8 (hardcover)
978-1-63739-328-4 (paperback)
978-1-63739-427-4 (ebook pdf)
978-1-63739-380-2 (hosted ebook)

Printed in the United States of America
Mankato, MN
082022

ABOUT THE AUTHOR

Susan Wroble is a children's writer with a passion for science, education, and dogs. She has degrees in electrical engineering and foreign affairs. When she isn't writing, you can find her working with her husband and their therapy dogs in Denver, Colorado.

TABLE OF CONTENTS

CHAPTER 1
Creating New Corals 5

CHAPTER 2
A Climate Emergency 9

CHAPTER 3
Climate Engineering 15

THAT'S AMAZING!
Carbon Capture 20

CHAPTER 4
The Fight Ahead 23

Focus on Climate Engineering • 30
Glossary • 31
To Learn More • 32
Index • 32

CHAPTER 1

CREATING NEW CORALS

Around the world, **coral reefs** are dying. **Climate change** is causing the oceans to be warmer. It is also leading to more severe storms. The crisis is increasing **ocean acidity** as well. As a result, many once-colorful reefs are now bleached white. Efforts to protect them have often failed.

Coral bleaching from climate change has made coral reefs some of the most at-risk environments on Earth.

An engineer at the Australian Institute of Marine Science tests a new coral that could survive warmer temperatures.

In response, a group of **engineers** came up with a new approach. These scientists worked at the Australian Institute of Marine Science. They worked in a research aquarium near Australia's Coral Sea. The main building is approximately half the size of a soccer field. Inside the building are dozens of tanks.

The seawater in each tank can be adjusted. Scientists can change the temperatures. They can also adjust the lighting and the food supply. That way, scientists can make a tank match the ocean as it is now. But they can also match the warmer ocean of the future.

Coral is an animal. Algae live inside it. Algae are tiny plant-like organisms. They give coral its bright colors. When water heats up, the coral pushes out the algae. Often, the coral then dies. At the research aquarium, scientists tested corals. They studied the algae. They were able to develop coral that can survive higher water temperatures.

CHAPTER 2

A CLIMATE EMERGENCY

In 1712, a scientist invented a steam engine. Powered by coal, the engine did the work of many people. Then inventors developed more machines that used **fossil fuels**. These fuels powered cars, trucks, and power plants. As humans burned these fuels, gases such as carbon dioxide (CO_2) entered the atmosphere.

In the early 2020s, more than 40 percent of the world's CO_2 emissions came from coal.

 In 2021, Hurricane Ida caused massive flooding in the northeastern part of the United States.

These gases trap heat. They are called greenhouse gases. They raise Earth's temperature and cause climate change.

A warming climate creates serious problems. Warmer air holds more water. Wetter air can cause stronger storms and more flooding.

Warm air causes ice to melt, too. This added water is especially serious near Earth's poles. Warmer water also takes up more space than cooler water. Together, these two changes raise sea levels. Rising seas and heavy storms damage the wetlands that protect the coast. They also threaten everyone living in coastal areas.

By the 1980s, scientists understood the problems of climate change. But governments did not pass enough laws to stop these problems. Energy companies continued to use natural gas and coal. Car companies built more and more gasoline-powered engines. The problem became an emergency.

Solar power produced approximately 1 percent of all the world's energy in the early 2020s.

Thankfully, scientists developed new sources of energy. For example, wind power and solar power produce no direct CO_2 **emissions**. By the 2020s, the technology to stop making new emissions was ready. Not adding emissions is known as net zero.

However, even that huge shift could not stop climate change. That's because people had already produced so much CO_2. And CO_2 stays in the atmosphere for many decades. As a result, people must also remove greenhouse gases. To do that, people need climate engineers.

REMOVING CO_2 IN SPACE

In 1970, three astronauts were in space. During their flight, an oxygen tank exploded. It damaged their spacecraft. The astronauts moved into a small lander. It was meant for only two people. As they exhaled, deadly CO_2 built up. Engineers found a solution. The astronauts used duct tape and socks. They changed the way their air filter worked. They showed that dangerous gases can be removed. It can even happen in space.

CHAPTER 3

CLIMATE ENGINEERING

Climate engineering involves very large projects where people change Earth. In some cases, people may make changes to the land. In other cases, they may change the ocean or the atmosphere. There are multiple approaches to climate engineering. Some take greenhouse gases out of the atmosphere. Other

Mangroves remove lots of CO_2. They also protect shores from rising sea levels. One climate engineering approach helps restore these areas.

climate engineering ideas aim to block sunlight.

Engineers tested many ways to reduce greenhouse gases. Some used filters to pull CO_2 from smokestacks. Others learned how to pull CO_2 directly out of the air. Still others tested adding iron to ocean water. The iron fed tiny plants and animals. These plants and animals collected carbon. When they died, they sank to the ocean floor.

Engineers also knew that plants store carbon. The engineers heated plants in a process called **pyrolysis**. It created small pieces called biochar. These pieces contain carbon. Putting biochar in the ground stores carbon safely. It also improves the soil for crops.

In addition, engineers found ways to block sunlight. They worked on ways to send some of the sun's energy back

PYROLYSIS AND BIOCHAR

Using pyrolysis can make fuels and biochar from plants. The biochar can go in the ground. Vehicles can use the fuel.

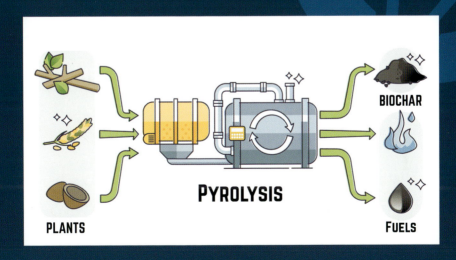

into space. This lowers the planet's temperature. Brighter surfaces reflect light and heat. Roads and buildings were repainted in light colors. In the oceans, engineers tested machines that blew bubbles. Bubbles make water brighter.

MOUNT TAMBORA

Putting small particles in Earth's atmosphere can lower temperatures. Scientists know this because of volcanoes. In 1815, Mount Tambora erupted in Indonesia. Ash from the volcano blocked sunlight for months. As a result, temperatures dropped around the world. Crops died. Engineers do not want these harmful effects to happen. But studying volcanoes was useful. It showed one possible way to help with climate change.

Greenland's Jakobshavn Glacier is melting fast. Walls could slow its ice loss. That could help slow sea-level rise.

In the air, engineers researched putting chalk dust into clouds. The dust would act like tiny mirrors to reflect sunlight.

Engineers also addressed the effects of climate change. They studied how to slow rising sea levels. One idea was to build walls near glaciers. The walls could keep warmer water from touching and melting the glaciers.

CARBON CAPTURE

Many companies collect CO_2 from factory smokestacks. But one company wanted to try something harder. The company wanted to capture carbon directly out of the air.

The company named its buildings Orca, after the whale. In its eight buildings, fans push air through special filters. In 2021, the filters began collecting CO_2. The plant captured nearly 4,400 tons (4,000 metric tons) in its first year.

Suddenly, the company had lots of carbon. This created a new problem. The company needed to store it safely. It had to stop the carbon from going back into the air. Another company had an answer. Machines mixed the captured gas with water. It became like a giant soda stream. The fizzy water was pumped underground. Then it was

Orca was the world's largest carbon-capture plant when it opened in 2021.

released. The gas reacted with minerals. In less than two years, the carbon should turn into stone. It will stay underground for millions of years.

CHAPTER 4

THE FIGHT AHEAD

Many people are fighting climate change. They drive electric vehicles. Or they walk or bike instead of traveling by car. These actions all help, but much more is needed. If nothing changes, global temperatures will climb even higher. The most important step to stop the warming is to reduce emissions to

One idea to lower Earth's temperature involves making cirrus clouds thinner. Cirrus clouds are wispy and form high in the atmosphere.

zero. The next step is to get rid of the CO_2 already in the atmosphere.

However, in the early 2020s, many engineering solutions were still in early stages. For example, carbon-capture plants remained expensive. Also, the plants were not removing enough CO_2. In 2021, plants removed only 14,300 tons (13,000 metric tons) of CO_2. That's less CO_2 than 3,000 cars produce in a year. Meanwhile, more than one billion cars were on the world's roads.

Storing carbon was another challenge. Engineers tested ways of storing carbon in rocks, plants, soils, and even the ocean floor. But the types of rocks that can store

Canada is home to the world's first carbon-capture project on a coal plant. It began capturing CO_2 in 2014.

carbon are found only in some places. Plus, forest fires can release carbon stored in plants or soil. And nothing stays on the ocean floor forever.

Getting carbon out of the atmosphere will take time. That is where solar engineering comes in. Putting particles into the air can lower the temperature. But is not a perfect solution. It may change rainfall patterns. Some areas might not receive enough rain. Others

AEROSOLS IN THE AIR

Spraying tiny particles called aerosols into the air could help lower Earth's temperature. The aerosols would reflect the sun's heat.

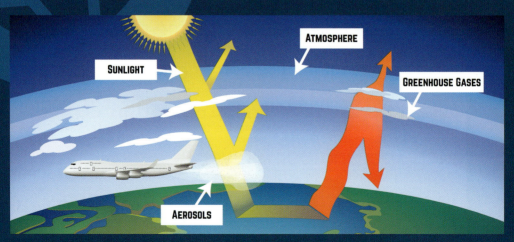

might be flooded. That would harm the food supply. In addition, the particles would need to be replaced often. Governments have not decided who would pay for that.

Solar engineering also does not solve other climate crises. It does not lower ocean acidity. It does not fix rising sea levels. More than 200 million people live in low-lying areas. As waters rise, these people will be forced to move.

Engineers kept working on solutions. Some developed ways to reduce runoff. One idea was asphalt that lets rainwater seep through. Another was putting gardens on the roofs of buildings. These solutions can reduce flooding. They do not stop sea levels from rising. But they can help people adapt to the impacts.

Part of the solution to global warming is reducing emissions to zero. But people must also remove the carbon that's

FLYING WITH BIOFUELS

In the early 2020s, most planes used fossil fuels to fly. But some flights were using fuels made from plants. These fuels are called biofuels. They can be made from food waste, cooking oils, sugarcane, and more. Burning biofuel releases CO_2 into the air. But it doesn't release any more CO_2 than it removed. That's because plants take in CO_2 as they grow. As a result, the fuel is **carbon neutral**. Biofuel-powered flights could be one step toward zero emissions.

Engineering cities to adapt to climate change is one important step for the future.

already in the air. Climate engineering has many risks. But it may also give people time to develop new technologies. To fight climate change, engineers know that there is no single answer. They will need many technologies.

FOCUS ON
CLIMATE ENGINEERING

Write your answers on a separate piece of paper.

1. Write a paragraph describing the main ideas of Chapter 1.

2. Do you think climate engineering is needed for our planet? Why or why not?

3. What is the goal of solar engineering?
   - **A.** creating energy without making greenhouse gases
   - **B.** taking greenhouse gases out of the atmosphere
   - **C.** lowering Earth's temperature

4. Why is stopping new CO_2 emissions not enough to stop climate change?
   - **A.** Too much CO_2 is already in the atmosphere.
   - **B.** CO_2 does not stay in the atmosphere for very long.
   - **C.** CO_2 is not an important cause of climate change.

Answer key on page 32.

GLOSSARY

carbon neutral
When something releases the same amount of carbon dioxide as it absorbs.

climate change
A human-caused global crisis involving long-term changes in Earth's temperature and weather patterns.

coral reefs
Systems of animals that live in warm, shallow waters.

emissions
Chemicals or substances that are released into the air, especially ones that harm the environment.

engineers
People who use science and math to solve problems.

fossil fuels
Energy sources that come from the remains of plants and animals that died long ago.

ocean acidity
A human-caused change in the chemical properties of ocean water that causes coral skeletons and animal shells to weaken.

pyrolysis
A process similar to burning but without using oxygen.

TO LEARN MORE

BOOKS

McCarthy, Cecilia Pinto. *Capturing Carbon with Fake Trees*. Minneapolis: Abdo Publishing, 2020.

Minoglio, Andrea. *Our World Out of Balance: Understanding Climate Change and What We Can Do*. San Francisco: Blue Dot Kids Press, 2021.

Santos, Rita, editor. *Geoengineering: Counteracting Climate Change*. New York: Greenhaven Publishing, 2019.

NOTE TO EDUCATORS

Visit **www.focusreaders.com** to find lesson plans, activities, links, and other resources related to this title.

INDEX

aerosols, 26
Australian Institute of Marine Science, 6–7

biochar, 16–17
biofuels, 28

carbon capture, 20–21, 24
carbon storage, 16–17, 20–21, 24–25

clouds, 19
coral, 5, 7

emissions, 12, 23, 28

glaciers, 19

Mount Tambora, 18

ocean acidity, 5, 27
ocean floor, 16, 24–25

runoff, 27

sea levels, 11, 19, 27–28
smokestacks, 16, 20
solar engineering, 25–27
steam engine, 9
storms, 5, 10–11

Answer Key: **1.** Answers will vary; **2.** Answers will vary; **3.** C; **4.** A